Detroit Warehouse Fire
Claims Three Firefighters
Detroit, Michigan

Investigated by: J. Gordon Routley

This is Report 003 of the Major Fires Investigation Project conducted by TriData Corporation under contract EMW-88-C-2277 to the United States Fire Administration, Federal Emergency Management Agency.

Homeland Security

Department of Homeland Security
United States Fire Administration
National Fire Data Center

U.S. Fire Administration Fire Investigations Program

The U.S. Fire Administration develops reports on selected major fires throughout the country. The fires usually involve multiple deaths or a large loss of property. But the primary criterion for deciding to do a report is whether it will result in significant "lessons learned." In some cases these lessons bring to light new knowledge about fire--the effect of building construction or contents, human behavior in fire, etc. In other cases, the lessons are not new but are serious enough to highlight once again, with yet another fire tragedy report. In some cases, special reports are developed to discuss events, drills, or new technologies which are of interest to the fire service.

The reports are sent to fire magazines and are distributed at National and Regional fire meetings. The International Association of Fire Chiefs assists the USFA in disseminating the findings throughout the fire service. On a continuing basis the reports are available on request from the USFA; announcements of their availability are published widely in fire journals and newsletters.

This body of work provides detailed information on the nature of the fire problem for policymakers who must decide on allocations of resources between fire and other pressing problems, and within the fire service to improve codes and code enforcement, training, public fire education, building technology, and other related areas.

The Fire Administration, which has no regulatory authority, sends an experienced fire investigator into a community after a major incident only after having conferred with the local fire authorities to insure that the assistance and presence of the USFA would be supportive and would in no way interfere with any review of the incident they are themselves conducting. The intent is not to arrive during the event or even immediately after, but rather after the dust settles, so that a complete and objective review of all the important aspects of the incident can be made. Local authorities review the USFA's report while it is in draft. The USFA investigator or team is available to local authorities should they wish to request technical assistance for their own investigation.

For additional copies of this report write to the U.S. Fire Administration, 16825 South Seton Avenue, Emmitsburg, Maryland 21727. The report is available on the Administration's Web site at http://www.usfa.dhs.gov/

U.S. Fire Administration

Mission Statement

As an entity of the Department of Homeland Security, the mission of the USFA is to reduce life and economic losses due to fire and related emergencies, through leadership, advocacy, coordination, and support. We serve the Nation independently, in coordination with other Federal agencies, and in partnership with fire protection and emergency service communities. With a commitment to excellence, we provide public education, training, technology, and data initiatives.

TABLE OF CONTENTS

DETROIT WAREHOUSE FIRE
CLAIMS THREE FIREFIGHTERS

A massive fire destroyed two large warehouse complexes and resulted in the deaths of three members of the Detroit Fire Department on March 12, 1987. The circumstances were particularly unusual, in the fact that the fatalities resulted from two separate incidents, in different buildings, almost three hours apart. The manner in which the fire conditions changed very rapidly, resulting in one fatality and several injuries, is also unusual and could easily have resulted in several additional firefighter deaths.

The Detroit Fire Department has determined that this fire resulted from an act of arson and a suspect has been charged with three counts of homicide. This report does not address the cause and origin aspects of the investigation. It is intended to describe the situation that was faced by the forces involved in fire suppression operations and the circumstances that resulted in the fatalities.

This report is for informational purposes only.

The fire scene was approximately two miles northwest of the downtown business area in a warehouse district one block from the Detroit Fire Department Training Academy. It involved two adjacent warehouse complexes, each constructed in several phases between approximately 1900 and 1920. The buildings in the north complex were abandoned, while the south complex was occupied by an operating paper products distributing company. The two complexes were separated by an entrance alley, approximately 20 feet wide that provided entry to a large yard area. This was the only access to the yard with the other sides blocked by buildings, railroad tracks, and retaining walls.

NORTH COMPLEX

The north complex consisted of an L-shaped three-story warehouse with full basement, facing the frontage road of I-96, attached to a four-story warehouse on the northeast corner of the complex. A fenced-in vacant lot on the northwest corner of the property was covered with old tires.

The three-story buildings were made of heavy timber construction with numerous vertical openings for freight elevators, conveyors, and stairs. The front section was 190 feet by 90 feet, divided down the middle by a firewall. The section extending east was 300 feet by 50 feet, with an intermediate firewall at the mid-point. Many of the fire doors were missing or damaged, negating any effective horizontal or vertical separations in this part of the complex. The four-story section was 230 feet by 100 feet, constructed of reinforced poured-in place concrete. The buildings were separated by a covered shipping area, approximately 20 feet wide and spanned by three crossover bridges. A railroad siding on top of a concrete viaduct ran along the south side of the three-story section at the second floor level.

The buildings had been protected by automatic sprinklers but the system was inoperative and had been partially dismantled, including removal of the sprinkler heads and the elevated storage tank.

The three-story section was formerly occupied by a wiping cloth distributor and was left heavily stocked when the company went out of business. The contents included rags in bales and crates, in

1

addition to piles of discarded clothing and materials. The bales were described as 5 feet in diameter and up to 8 feet tall, bound in burlap. Some of these rags may have been oil soaked, adding to the intensity of the fire and rapid fire spread.

The four-story building was previously used by a division of the same company that dealt in used rags and other bulk items. It was also abandoned, fully stocked with combustible contents. Efforts to keep the abandoned property secured had been unsuccessful and transients were known to frequently occupy the buildings.

Fire companies in the area were aware of the risk and had conducted several familiarization tours of the buildings. During these tours, the hazards of open passages between floors had been noted and firefighters had moved contents to cover several floor openings they considered hazardous. At least one fire had occurred previously in the buildings and firefighters anticipated another, sooner or later. Many of the personnel responding on the first alarm had toured the building in the previous month. The property had been abandoned in 1982 and came under State control for back taxes. The City of Detroit had awarded a contract for its demolition and the contractor was scheduled to begin work within a few days after the fire.

SOUTH COMPLEX

The south complex was occupied by a paper products distribution company. This section included three-story warehouse sections and additional sections of one and two stories, forming a triangle with frontages of 320 feet along the I-96 access road, 300 feet facing the yard area, and 400 feet along the railroad embankment to the south. The railroad tracks coincided with the second floor level on the south side.

The front of the paper company's buildings had been covered with a metal façade, obscuring the age and complexity of the structures. But these factors were plainly visible from the sides and rear. The construction included steel frame and heavy timber sections, divided by several firewalls and protected by automatic sprinklers. The contents included paper and plastic products, including packing materials and consumer goods.

Examination of the scene after the fire revealed that the brick construction in both complexes was of inferior quality, presenting a high risk for early collapse. The brick work included many irregularities and no visible reinforcing with evident gaps between the double and triple courses of brick. The construction had taken place over several years resulting in complicated arrangements of buildings and many construction features that could not be seen from the exterior.

The fire was reported in the abandoned section of the complex at 1506 hours on March 12, 1987. A first alarm assignment consisting of three engine companies, one ladder company, a squad, and battalion chief was dispatched and the first due ladder company arrived within two minutes. Ladder 9, responding from its quarters at the Training Academy, one block away, had a view of the north and east faces of the buildings as they approached. They arrived, reporting a small amount of light smoke showing from the southeast corner third floor windows.

The first arriving companies had to force entry through the front doors and then make their way via an unenclosed interior stairway to the top floor. At this level they found at least two small fires in trash and rags toward the south east of the floor area. They went to a window and dropped a rope, intending to pull a 1-1/2-inch hoseline up to the third floor. The fire did not appear to be threatening at this point and the crews anticipated a quick and easy job of extinguishment.

Very suddenly, the conditions on the third floor changed dramatically. A heavy front of smoke and flame rolled over on the interior crews, forcing them to abandon their positions and retreat toward the stairs. One firefighter, looking through a doorway to the adjoining sections, reported a mass of flames approaching rapidly. There are indications that an additional fire, possibly set on a lower floor toward the middle of the east-west wing, had reached the flashover stage and was rapidly engulfing the entire third floor. A total of eight firefighters were on the third floor when the flashover occurred, forcing them to crawl back toward the stairway. Two managed to dive down the stairs and escape with hand and facial burns and other injuries. One went out a window to an aerial ladder but five could not reach the stairs and were trapped by the flashover.

The firefighters found their way to windows in the northwest quadrant of the building and called for help. One lieutenant lost his grip while hanging out a window and fell, striking a ledge at the second floor level and landing head first on the street below. Other firefighters initiated CPR and transported him in a fire investigator's sedan to a hospital where he was pronounced dead on arrival.

While a ground ladder was being raised, a second firefighter fell, striking a large telephone line and landing on the street with a fractured elbow and shoulder. Two were rescued with an extension ladder from a front window, while a master stream was used to protect another, hanging onto the sill of a window around the corner. To reach him, a short ladder had to be used to scale a chain link fence into the vacant yard and a 45 feet extension ladder was passed over and raised.

All of the firefighters were wearing full protective clothing, including coats and helmets that complied with NFPA standards, leather gloves, and 3/4-length rubber boots. Most wore Self-Contained Breathing Apparatus (SCBAs) on their backs and four were able to don their facepieces as they crawled toward the stairs, including the two who fell. Several of the firefighters received first and second degree burns to their hands, wrists, thighs, and necks.

EXPOSURE PRIORITY

When conditions began to deteriorate, the battalion chief immediately called for a second alarm at 1519 hours. This was followed by a third alarm at 1522, a fourth alarm at 1530, and a fifth alarm at 1538 hours. Requests for additional companies escalated the response to the equivalent of seven alarms as flames rapidly engulfed the entire three story building. The deputy fire chief, responding on the third alarm, assumed control and attempted to deploy companies to confine the fire. A ladder company, assigned to set up a master stream at the east end (Lawton Street) of the building could not operate in an effective position because of the imminent danger of wall collapse, while companies could not use the alley entrance to the yard because of the collapsing wall at that end of the fire building. A railroad viaduct at the second floor level severely restricted access on the east side of the fire building and handlines had to be stretched by hand into the yard area over a retaining wall and fence, to apply water on the south side.

As the flames raged on the heavy fuel load of the rag warehouse, the exposed paper warehouse was quickly ignited. Moments after the rescue of the trapped firefighters was completed, flames were visible under the eaves and in the exposed windows at the alley opening. Companies were sent to the roof and interior to attempt to stop extension with handlines, while elevated master streams were set up to cover the exposure from the front, including a tower ladder set up on the freeway off the ramp in front of the complex.

This holding action was successful in the front part of the building, but flames soon overwhelmed the sprinkler system and broke through the roof of the next section to the east. The portions of the warehouse that were beyond the reach of elevated streams received a heavy exposure from radiant heat, flying brands and internal extension, resulting in full involvement of the northeast section of the paper products warehouse. Handlines and portable master stream devices were moved onto the railroad right-of-way to attack the growing fire and the deputy chief ordered all personnel out of the warehouse because of the danger of structural collapse.

With the three-story warehouses now fully involved around three sides of the yard, the operation went into a long duration defensive mode. The unusually high fuel load in both complexes created intense thermal columns and showered the area to the east with flying brands. Several small fires were handled by engine companies and citizens with garden hoses, up to three blocks downwind.

STRUCTURAL COLLAPSE

Approximately two hours into the defensive battle, companies working along the east side of the paper warehouse began to approach the burning three-story section over the roof tops of the uninvolved two-story sections. Near the middle of the building they found an area where the roof had burned off and the third floor contents were partially burned. It is believed that the fire in this area had been controlled by an elevated stream, operated from the front of the building. This section was separated by a firewall from the fully involved section to the east.

Members of two companies made their way from the adjoining rooftop into the third floor area where they picked up and extended handlines that had been abandoned earlier. Three members of one company were working near a firewall, overhauling debris, when a section of parapet collapsed without warning at 1758 hours. This wall was free standing at the time, since roof structures on both sides had burned away and an intense fire was burning on the other side.

Two members were caught by the falling wall, which collapsed the third and second floors down into the first floor and were trapped in a pile of debris approximately 1.2 feet deep in the unburned ground floor area. The remaining member of the company called for help and made his way downstairs to begin digging for the victims. He was joined by numerous other firefighters as the word of men trapped was announced over the fireground radio channel.

The deputy fire chief organized a rescue effort with crews digging out bricks and debris by hand. This effort actually took place inside the ground floor warehouse area and efforts to limit the number of personnel exposed to the danger of further collapse were hampered by the urgency to rescue trapped comrades. While this operation was in progress, the application of water was severely curtailed to avoid causing further collapse of the structure. An additional assignment of fresh companies was called to assist in the rescue operation.

The rescue effort took more than one hour, resulting in the recovery of the bodies of a 58-year-old lieutenant and 20-year-old probationary firefighter. At the time of the body recovery, the deputy fire chief was contemplating removing all crews because of deteriorating fire conditions, and as soon as the operation was completed the building was abandoned.

By this time the fire had extended into most of the two-story sections of the paper warehouse and total destruction was unavoidable. All crews withdrew to safe positions and master stream operations were continued for almost 24 hours. Late on the following afternoon, the contractor who held the demolition contract was called into level the unsafe walls of both complexes.

LESSONS LEARNED

1. The buildings involved in this fire were heavily loaded with highly combustible contents. The arrangement of the buildings created conditions that were ripe for very rapid fire growth and spread, particularly in the building of fire origin where the automatic sprinkler system had been rendered inoperative and fire doors and walls were compromised. These factors must be noted during pre-fire planning visits and responding companies must be prepared to deal with extremely rapid fire growth conditions.

2. Vacant buildings often present an attractive nuisance to members of society who engage in the crime of arson, either for profit or for more unpredictable motives. Where these problems exist, companies should make a priority of pre-fire planning and familiarize themselves with access, contents, special hazards, and hidden traps that may be critical in a firefighting operation. Efforts must also be directed toward having abandoned buildings secured or demolished as quickly as possible. The actions of an arsonist are truly unpredictable. Firefighters encountering obvious arson situations must be particularly vigilant for multiple points of origin, accelerants, and other factors that could cause rapid changes in fire conditions.

3. The contents of the rag warehouse were not only highly combustible, but also obstructed access through the storage areas and prevented firefighters from finding the stairway when the unexpected flashover occurred. When encountering conditions of this nature, the use of guide ropes or hoselines to lead the way back to an exit should be seriously considered.

4. When fighting any type of fire, the officer in command must consistently be aware of the risk to personnel and question whether the potential results justify the risk. No building is worth the life of a firefighter and abandoned or substantially destroyed buildings do not justify risking personnel under any circumstances. This must be weighed in context. However, as it is not realistic to allow any fire in an abandoned building to burn unchecked. If manageable fires are not controlled in their early stages, they will inevitably grow to major proportions and create a much greater risk to firefighters and the community.

5. On their arrival, firefighters encountered a situation that did not appear to present any significant danger. When the flashover occurred, they were very suddenly subjected to intense heat and flames, as well as zero-visibility smoke conditions. Those members who had their SCBAs on their backs had time to don their facepieces as they crawled from the danger. Those who did not have SCBAs, at least on their backs, were suddenly in extreme danger and were lucky to survive.

    Whatever the situation, company officers must take basic precautions, including:

    *   Identifying secondary means of escape;

    *   Ensuring that all personnel are wearing full protective clothing and SCBAs;

    *   Maintaining accountability for all crew members at all times;

    *   Constantly being aware of their surroundings and changing conditions; and,

    *   Being trained to react to unanticipated emergency conditions.

6. Large scale operations, such as this one, require strong centralized command to establish and communicate the basic strategy that will be employed and to coordinate operations. This must be supported by a fireground organization that controls the tactical position and function of all operating units and monitors safety conditions, in accordance with the strategic plan. This requires sufficient command level officers and effective communications to perform the essential tasks. Offensive and defensive firefighting operations must never be mixed or confused. When a fire is being managed in a defensive mode, all personnel must be aware of the strategic plan and stay out of the uninvolved area until re-entry is authorized by the officer in command of the incident. "Free lance" operations must not be allowed.

7. The inherent structural weaknesses of a building may or may not be plainly visible. Potential weak points should be noted during pre-fire planning and inspection visits. Safety officers should be assigned to monitor the operation and to evaluate conditions as they are encountered all personnel should be able to recognize signs of weakness or impeding collapse.

8. The urgency of a rescue operation may cause firefighters to accept severe risks and to endanger themselves in large numbers. The risk of additional losses must be weighed against the changes of a successful rescue. The loss of control may escalate a tragic situation into a major disaster.

The "lessons learned" listed in this report should not be interpreted as criticism of the Detroit Fire Department or of any of the individuals involved in this incident. This fire presented an extremely unusual combination of circumstances that had a devastating effect on a very competent and experienced fire department. The incident involved heroic actions and tragic results that were felt throughout the Detroit Fire Department and throughout the fire service. These lessons should be taken as reminders that firefighters must never relax their attention to basic safety procedures and must always be prepared to deal with a situation that changes from routine to critical without warning.

This report was prepared by J. Gordon Routley, Chairman of the Health and Safety Committee of the International Association of Fire Chiefs and by TriData Corporation for the United States Fire Administration. The cooperation of the Phoenix Fire Department in making Mr. Routley available for this project is acknowledged.

In addition, the cooperation of Commissioner Melvin Jefferson and the members of the Detroit Fire Department is greatly appreciated.

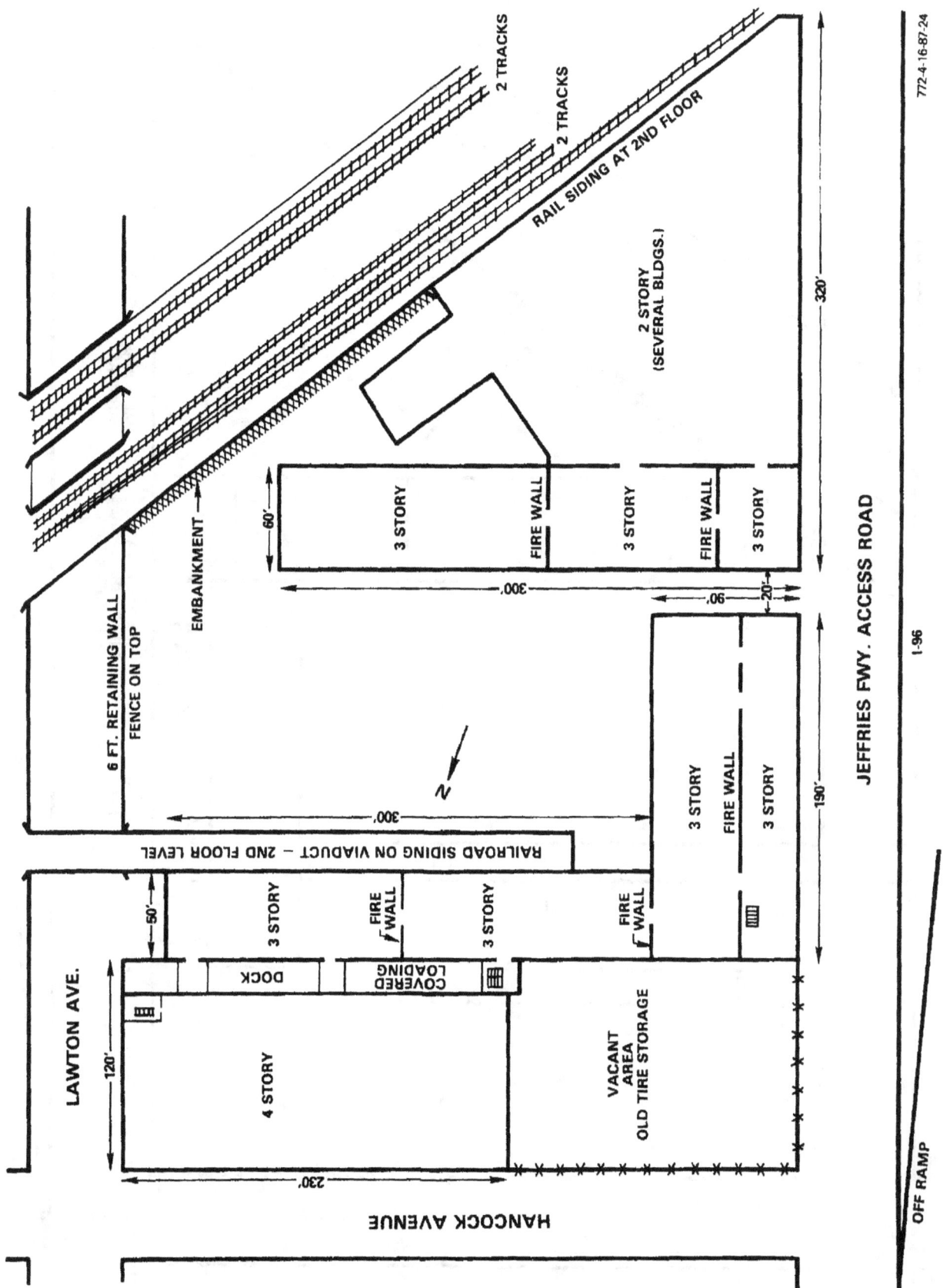

772-4-16-87-24

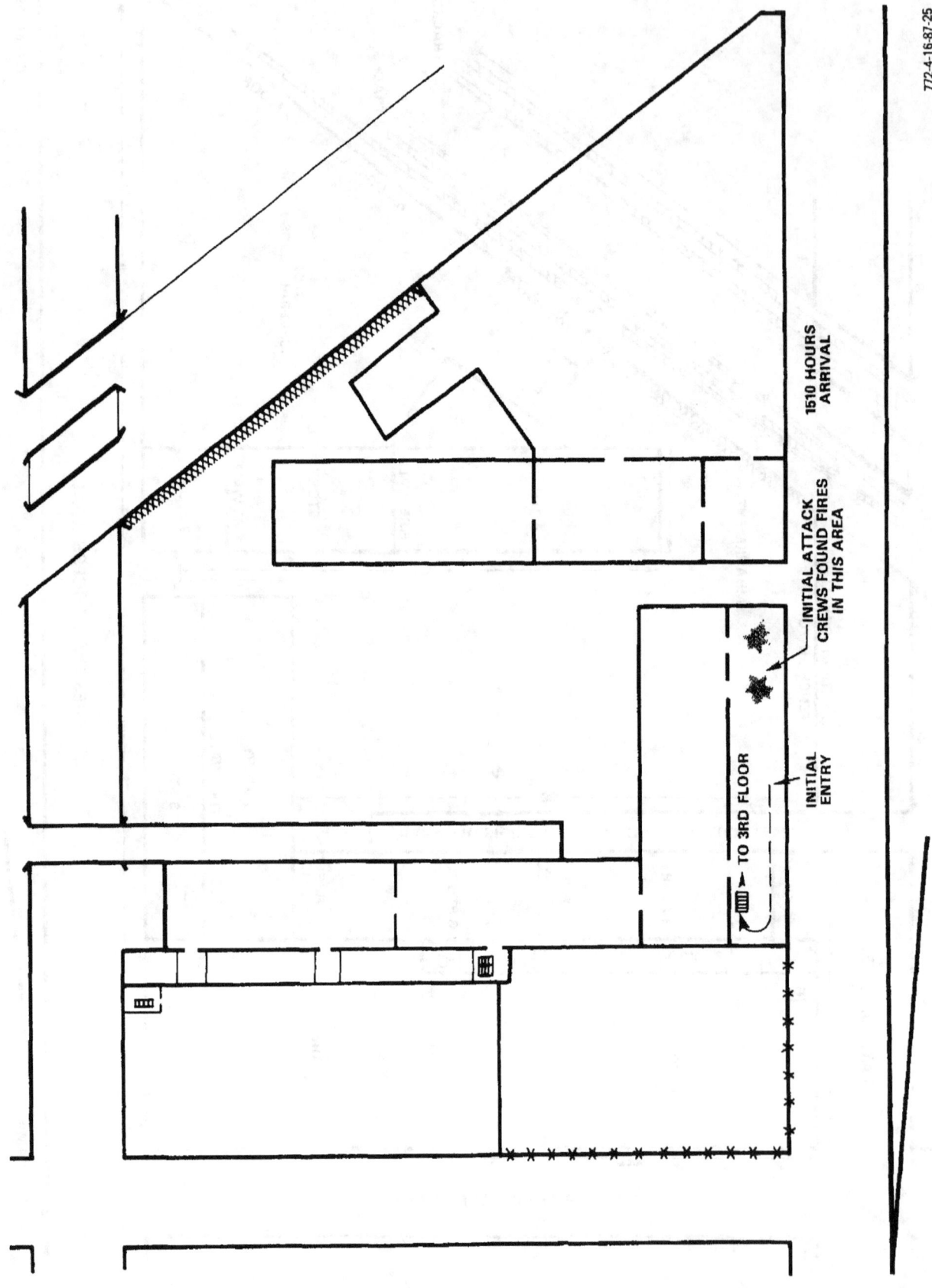

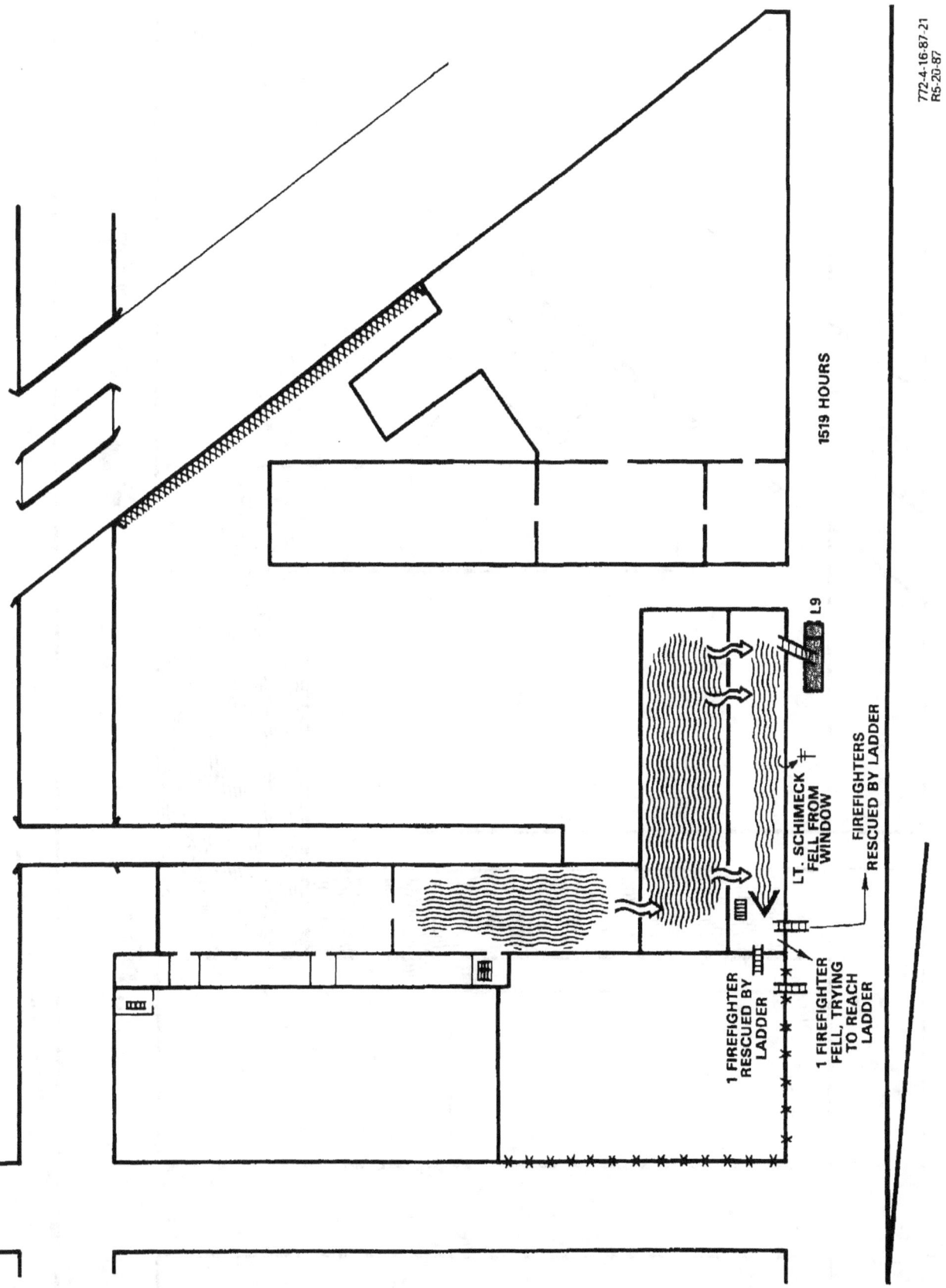

1519 HOURS

LT. SCHIMECK FELL FROM WINDOW

FIREFIGHTERS RESCUED BY LADDER

L9

1 FIREFIGHTER RESCUED BY LADDER

1 FIREFIGHTER FELL, TRYING TO REACH LADDER

772-4-16-87-21
R5-20-87

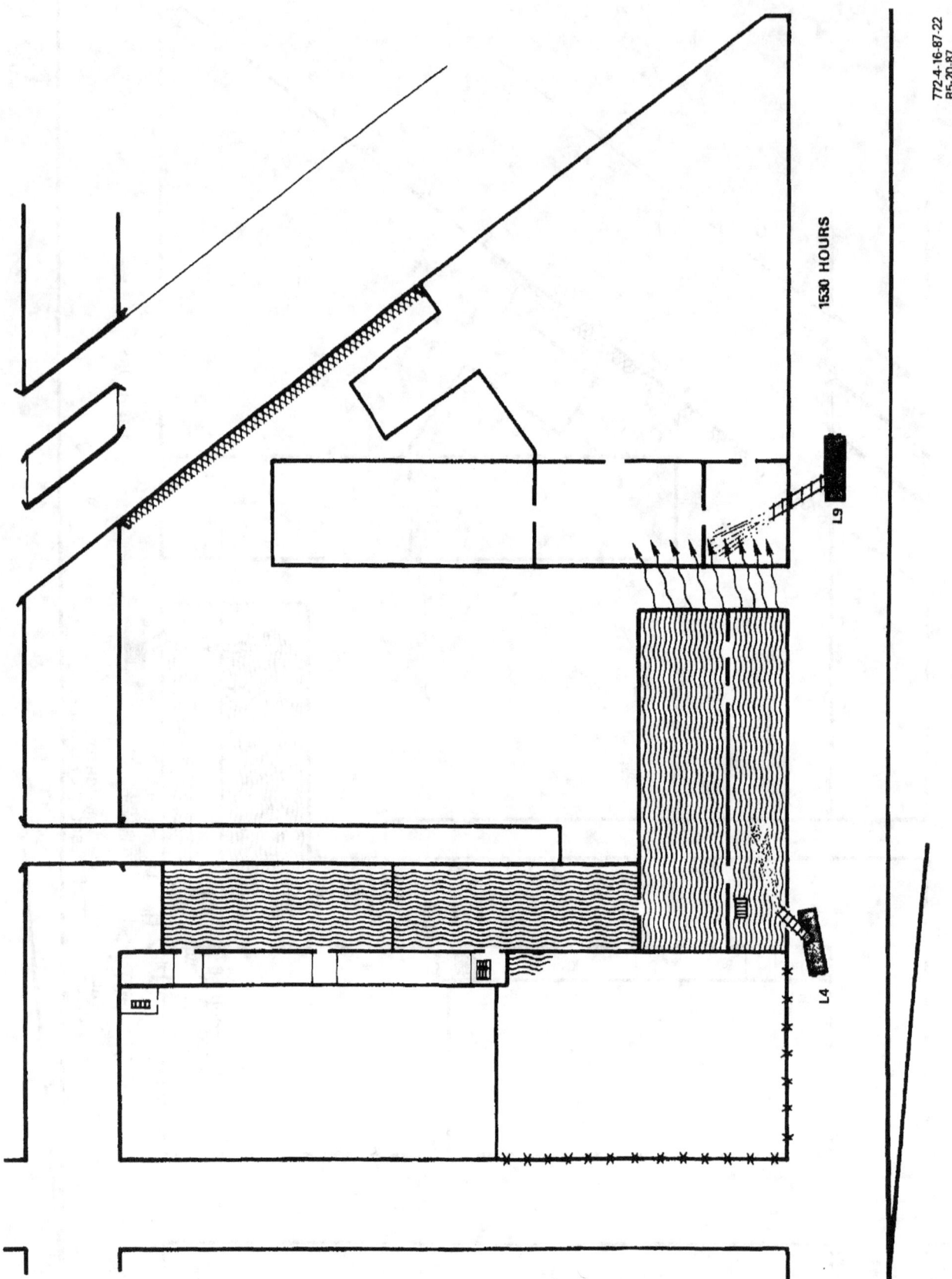

772-4-16-87-22
R5-20-87

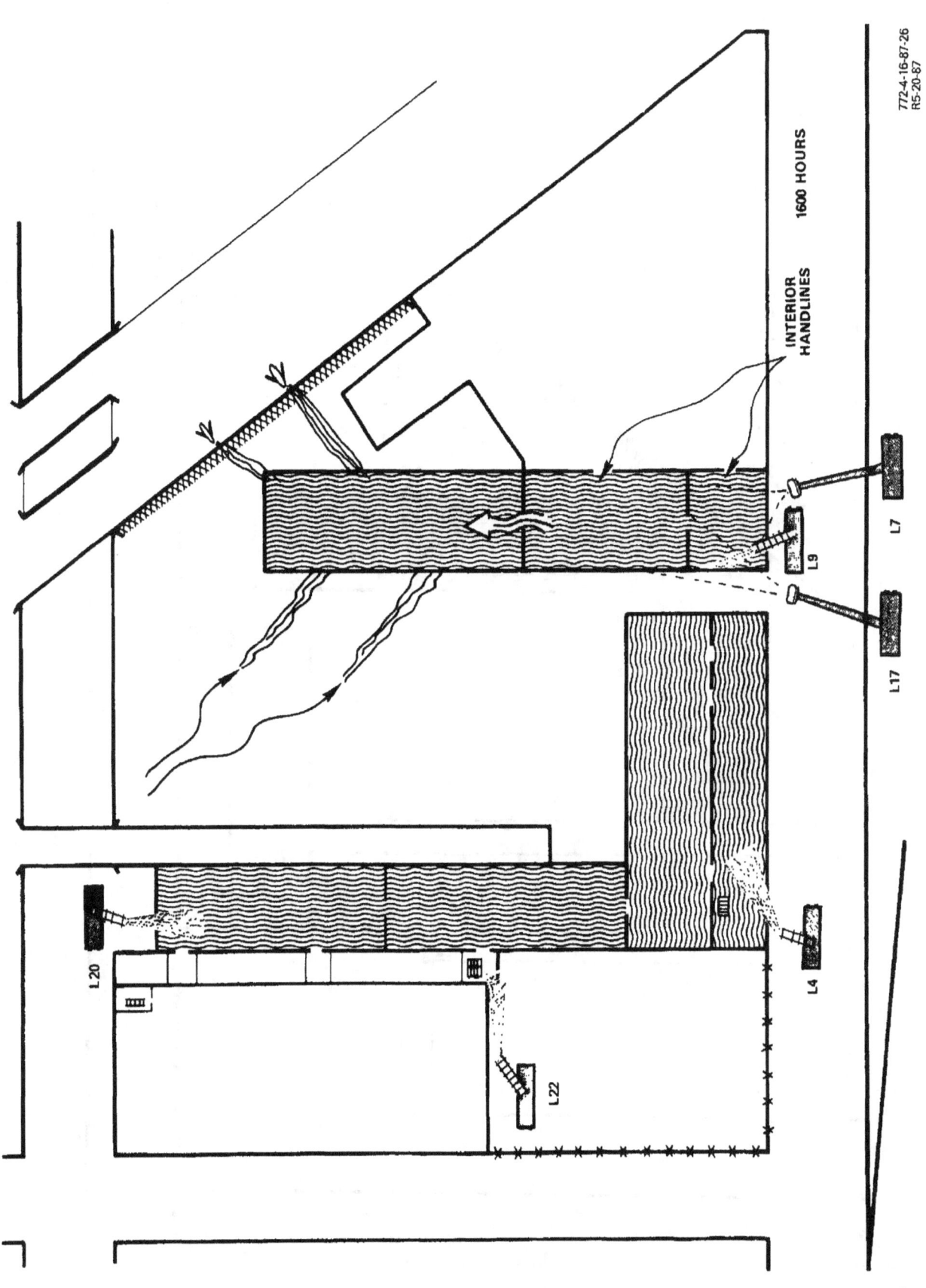

INTERIOR HANDLINES

1600 HOURS

L7

L9

L17

L20

L22

L4

772-4-16-87-26
R5-20-87

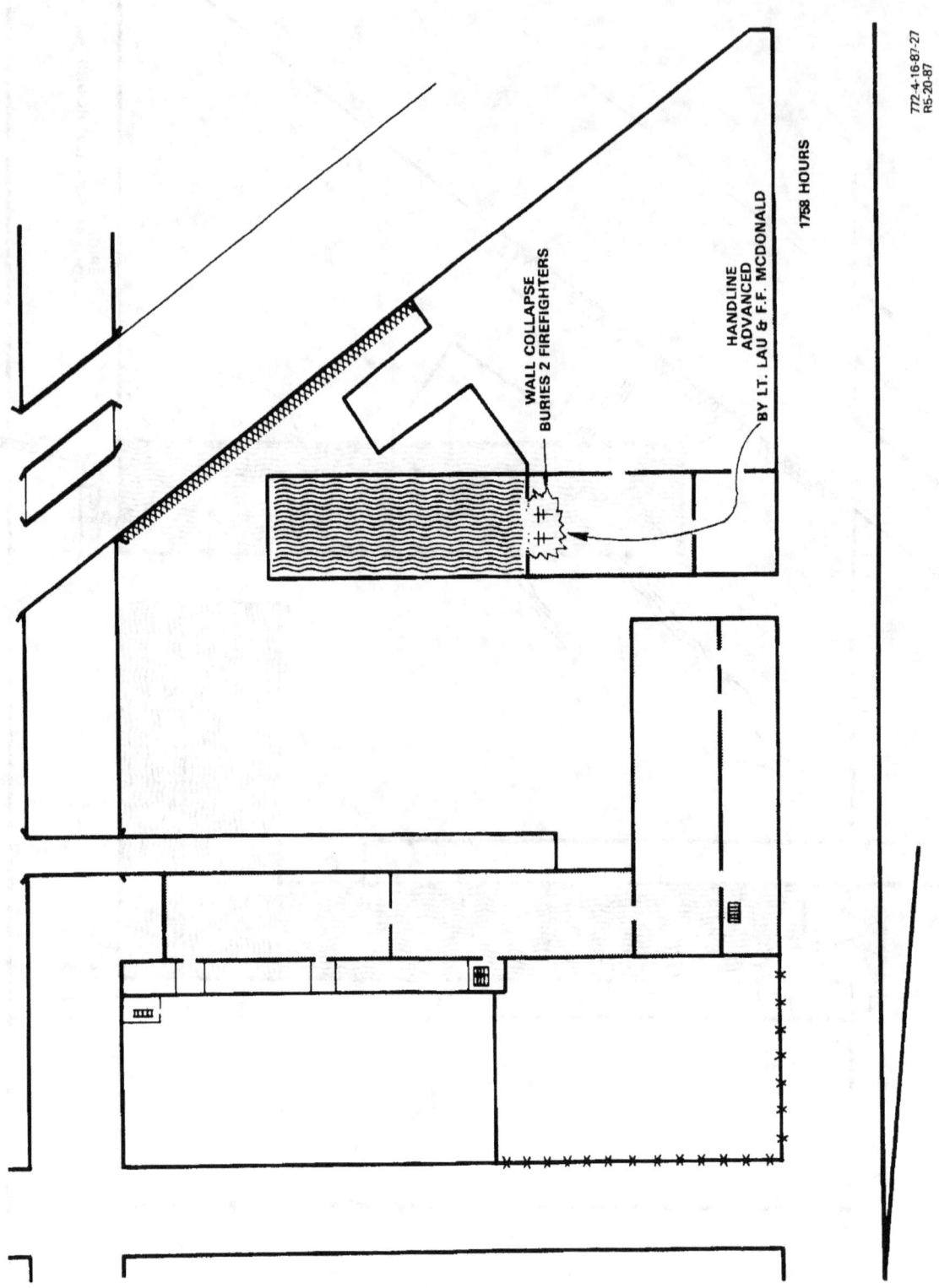

WALL COLLAPSE
BURIES 2 FIREFIGHTERS

HANDLINE
ADVANCED
BY LT. LAU & F.F. MCDONALD

1758 HOURS

772-4-16-87-27
RS-20-87

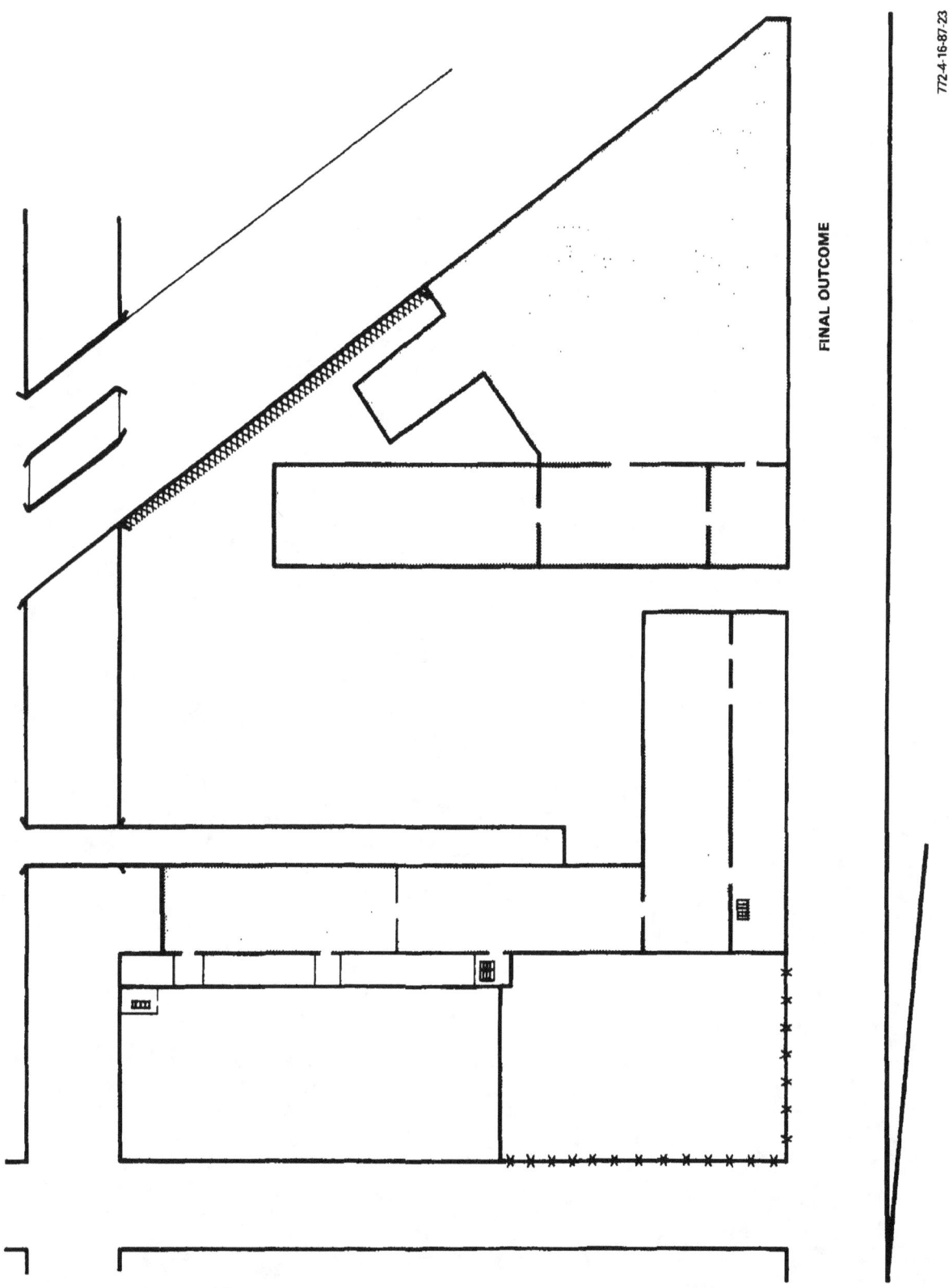

FINAL OUTCOME

772-4-16-87-23

PHOTO CAPTIONS

Photos 1-4

Aerial views of the incident scene – day after the fire.

NOTE: *Freeway is west exposure; railroad is south exposure.*

In photo #1, note Detroit Fire Department Training Academy – six-story building with radio tower, left of smoke plume.

Photos 5-7

Fireground activity near northwest corner of original fire building, shortly after flashover and rescue operations.

NOTE: *Ladder used to rescue trapped firefighter.*

PHOTO 5:   Master stream on Engine 10 used to protect firefighter in window prior to rescue. Also extension of flames to paper warehouse at eaves.

PHOTO 7:   Fire in tires at ground level and fire spread on second and third floors.

Photos 8-12

Continuing firefighting activity. Northwest part of the complex.

Photo 13

Rescue operations for trapped firefighters inside ground floor of paper warehouse. Both second and third floors have collapsed into unburned area.

Photo 1.

Photo 2.

Photo 3.

Photo 4.

Photo 5.

Photo 6.

Photo 7.

Photo 8.

Photo 9.

Photo 10.

Photo 11.

Photo 12.

Photo 13.